mountain
habitats

Written by Scarlett Lovell and Sue Smith
Wildlife illustrations by Robert Noreika
Cartoon illustrations by Miriam Katin

Chamois page 5

Grizzly bear cubs page 13

Tarsier page

Happy face spider page 20

Contents

Alps	page 3
Virunga Volcanoes	page 7
Canadian Rockies	page 11
Mount Kinabalu	page 15
Mount Wai'ale'ale	page 19
Glossary	page 23
Index	page 24

Flying lemur page 16

Mountain gorilla page 9

Alps

The Alps are famous for snow-covered peaks, extensive forests, and grassy meadows. The mountains form a great arc that covers more than 80,000 square miles (208,000 square kilometers) of south central Europe. In spring, snow melts from the lower elevations, but glaciers and snow remain higher up.

Alpine snowbells are the first spring flowers. Warm air pockets form under the snow and allow the leaves and stems to push through early.

stone pine

larch

rusty alpenrose

white edelweiss

Alpine snowbells

blue gentian

LET'S EXPLORE!

Golden eagle chicks flap their wings to develop their flight muscles.

A wall creeper uses its long bill to pick insects and spiders from deep cracks in the mountains.

An Alpine shrew finds its way by listening to echoes from ultrasonic sounds it sends out.

Alpine chough

Alpine ibex

crossbill

ptarmigan

The speckled brown feathers of the ptarmigan make it difficult for predators to see it among rocks and plants. In winter, it grows feathers that blend in with the snow.

A chamois can leap across mountain chasms 20 feet (6 meters) wide.

A weasel's long, thin body allows it to follow mice and moles into their tunnels.

Lynx kittens play-fight to develop the strength and quickness needed to be solitary hunters.

brown bear

lynx

red fox

mountain hare

Adaptation

As seasons change, some animals move from one part of the mountains to another to search for food.

Vertical Movement

MEADOW
During most of the year, chamois move up to the lush Alpine meadows and graze on grass and herbs.

FOREST
In winter, herds of up to 100 chamois go down to the forest to browse on conifer needles and bark.

CAN YOU FIND IT?

At a distance, the different kinds of vegetation all the way up the mountain look like the layers of a cake.

Turn back the page and find the Alpine marmot.

Virunga Volcanoes

The Virunga Volcanoes in central East Africa rise to over 14,800 feet (4,500 meters). Because the volcanoes are near the equator, it is very hot at the lower elevations. High rainfall creates dense rain forests where thousands of plants and animals live. On the upper slopes, it is colder.

In wet weather, lichen soak up water like a sponge, but when dry, lichen crumble to pieces if touched.

LET'S EXPLORE!

A duiker's splayed hooves keep it from sinking into the marshy ground.

A chameleon flicks out its tongue to capture a grasshopper.

A hyrax's feet act like suction cups, helping it cling to rocks.

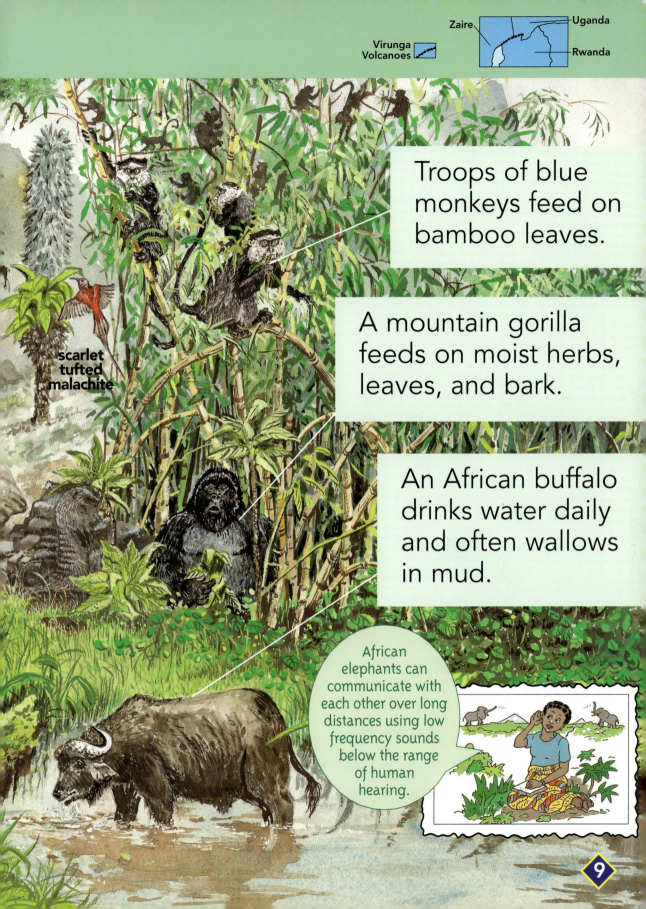

Adaptation

The diet of many animals, such as gorillas, varies with the season and the availability of food.

Seasonal Foods

BAMBOO FOREST
Bamboo trees produce new growth at the beginning of the rainy season. In October, the male leader of the gorillas takes the group into the forest to eat tender bamboo shoots.

WOODLAND
The bamboo forest is ignored during the rest of the year. The gorillas prefer to eat in the mountain woodland with its large variety and abundance of food plants.

In the wild, mountain gorillas do not drink water. They receive all the water they need from eating moist plants.

CAN YOU FIND IT?

Turn back the page and find the mongoose.

Canadian Rockies

The Canadian Rocky Mountains extend 1,500 miles (2,400 kilometers) along the length of Canada. The mountains are extremely rugged, with jagged, snow-capped peaks, turbulent rivers, and massive glaciers. Most of the trees are conifers, which can survive the long, cold, northern winters.

Because lodgepole pines are so straight, North American Indians used them to support their tipis and build their lodges.

LET'S EXPLORE!

Mountain goats stay safe from enemies by climbing steep slopes and jumping from rock to rock.

Gray wolves hunt in packs because they are more successful as a group.

A crossbill forces the scales of pine cones apart and lifts out seeds with its tongue.

Adaptation

The fur of many animals changes according to the season.

Fur Growth

WINTER
Mountain goats have coats of heavy white wool, which keep them warm.

SPRING
When temperatures are warmer, mountain goats shed their heavy coats.

WINTER
Snowshoe hares grow an extra thick coat of white fur, which blends in with the snow.

SPRING
Snowshoe hares molt and grow brown fur, which blends in with the forest.

Animals' fur may also change as they grow older. As cubs, mountain lions have spots on their fur, which make it more difficult for predators to see them in the forest.

CAN YOU FIND IT?

Turn back the page and find the golden mantled squirrel.

Mount Kinabalu

Mount Kinabalu in Southeast Asia rises 13,455 feet (4 kilometers) straight out of a tropical forest to a bare summit of granite. The warm lowlands are covered with colorful plants and flowers, while the summit is black rock and very cold. These differences in temperature often produce mist and rain.

Rafflesia flowers may be up to 39 inches (1 meter) wide. When open, they give off a putrid odor that attracts insects.

LET'S EXPLORE!

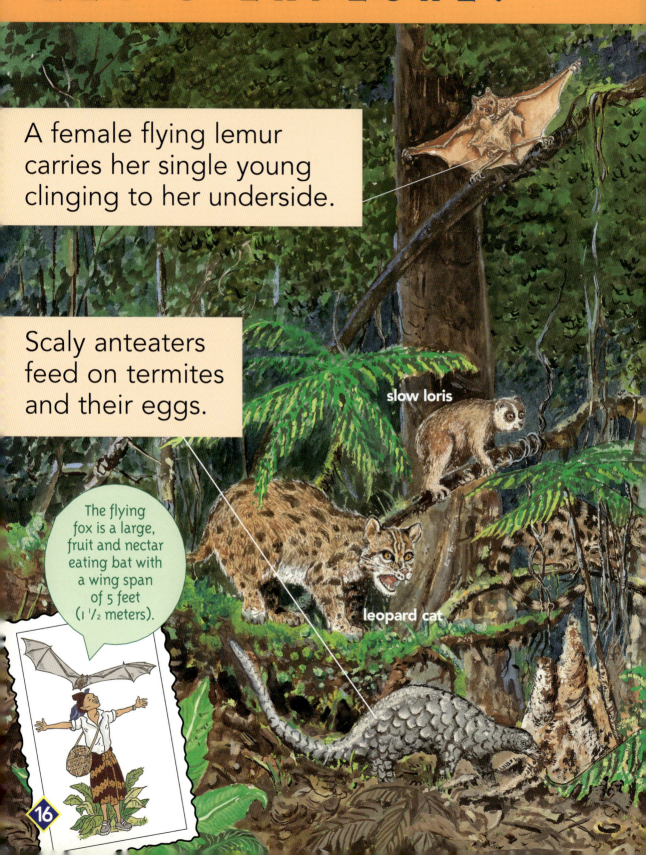

A female flying lemur carries her single young clinging to her underside.

Scaly anteaters feed on termites and their eggs.

slow loris

The flying fox is a large, fruit and nectar eating bat with a wing span of 5 feet (1 1/2 meters).

leopard cat

An orangutan sleeps in a tree on branches it pulls together to form a nest.

A tarsier's huge eyes help it see at night.

The Kinabalu shrew is one of the only mammals found at all elevations.

Adaptation

Some families of plants are able to grow in different areas of the tropical forest.

Growing Conditions

ON THE GROUND
Terrestrial orchids grow on the ground. They get nourishment from the soil.

ON TREES
Epiphytes grow on trees. Nutrients carried by the wind provide nourishment.

ON ROCKS
Lithophytes grow on rocks. They get nourishment from the air and rock particles.

There are over 25,000 species of orchids in the world. More than 1,200 of them grow on Mount Kinabalu.

CAN YOU FIND IT?

Turn back the page and find the banded linsang.

Mount Wai'ale'ale

The Hawaiian Islands are the most geographically isolated place in the world. Life was carried there by wind, water, and birds, and some species evolved that cannot be found anywhere else on Earth. On the island of Kaua'i, Mount Wai'ale'ale *(why-ah-lay-ah-lay)* rises 5,148 feet (1,560 meters).

Mount Wai'ale'ale is the wettest place on Earth. It averages about 480 inches (1,219 centimeters) of rain each year.

LET'S EXPLORE!

A grappler moth avoids capture because it has ears that enable it to hear.

An 'elepaio's nest of lichens and grass is held together with spider webs.

A happy face spider cares for its young and provides them with food.

Hawaiian land snail

black rat

The pinao, a dragonfly, is the largest native Hawaiian insect. Its wing span can be up to 6 inches (15 centimeters).

Pacific Ocean
Mount Wai'ale'ale ▲
Kaua'i

An 'a'o, or shearwater, lays just one egg each year.

cutaway view of nest

At night, the hoary bat uses sonar to find the insects it eats.

A grappler caterpillar is the only caterpillar that captures and eats insects.

'i'iwi

Kamehameha butterfly

sundew

pinao

Adaptation

Over time, the size and shape of honeycreepers' beaks varied as the birds adapted to a wide variety of available foods.

Beaks

'APAPANE
The slightly curved beak of the 'apapane allows it to feed on nectar and caterpillars.

'AKEPA
The short beak of the 'akepa is suited to searching for insects in flowers and on leaves.

'I'IWI
The long, curved beak of the 'i'iwi enables it to take nectar from tubular or bell-shaped flowers and capture fruit flies from other flowers.

NUKUPU'U
The shorter, lower beak of the nukupu'u is used for crushing and digging into wood and bark in search of insects to eat.

Some honeycreepers, like the nukupu'u, roll their tongues into tubes through which they can suck nectar.

CAN YOU FIND IT?

Turn back the page and find the gecko.

Glossary

adapt - to change in order to survive in new conditions.

chasm - a deep, wide crack in the earth's surface.

conifer - a tree that has cones, usually an evergreen.

elevation - the height to which something rises above sea level.

equator - an imaginary line around the middle of the earth that is an equal distance from the North Pole and South Pole.

evolve - to develop by gradual changes.

glacier - a huge mass of ice that moves very slowly down a slope or across land.

granite - a type of very hard rock.

herb - a plant whose stem above ground is not woody; some herbs are used in medicines or for flavoring foods.

lichen - a plant that looks like dry moss and grows in patches on rocks, trees, and other surfaces.

mammal - a warmblooded animal that usually has hair on its body; the female produces milk for its young.

marsh - low land that is wet and soft.

molt - to shed skin, feathers, hair, or a shell before getting a new covering.

nectar - a sweet liquid made by many flowers.

predator - an animal that lives by killing other animals for food.

prey - to hunt other animals for food. It can also mean the animal that is hunted.

putrid - rotten smelling.

solitary - living or being alone.

sonar - a way to detect and locate objects through the reflection of sound waves.

species - a group of plants or animals that are alike in certain ways and can breed together.

splayed - to be spread out or apart.

summit - the highest point or part of something.

tropical - in the area of the earth called the tropics; the climate is very hot.

ultrasonic - sound above the range of normal human hearing.

vegetation - all the plants or plant life of a place.

vertical - straight up and down.

Index

adaptation..........6, 10, 14, 18, 22
Alpine snowbell........................3
Alps................................3–6
anteater, scaly......................16
'a'o..................................21
bat, hoary...........................21
bamboo forest........................10
beaks................................22
bear, grizzly........................13
buffalo, African......................9
Canadian Rockies..................11–14
caterpillar, grappler................21
chameleon.............................8
chamois............................5, 6
cougar...............................13
crossbill............................12
duiker................................8
eagle, golden.........................4
'elepaio..........................20–21
elephant, African...................8–9
fox, flying.......................16–17
fur growth...........................14
growing conditions...................18
goat, mountain....................12, 14
gorilla, mountain..................9, 10
hare, snowshoe.......................14
honeycreeper
    'akepa............................22
    'apapane..........................22
    'i'iwi............................22
    nukupu'u..........................22
hyrax.................................8

Kaua'i, Hawaiian Islands.........19–22
lemur, flying........................16
lichen................................7
lion, mountain.......................14
lodgepole pine.......................11
lynx kittens..........................5
moth, grappler.......................20
monkey, blue..........................9
Mount Kinabalu....................15–18
Mount Wai'ale'ale.................19–22
orangutan............................17
orchid
    epiphyte..........................18
    lithophyte........................18
    terrestrial.......................18
osprey...............................13
pinao.............................20–21
ptarmigan.............................4
rafflesia............................15
seasonal foods.......................10
shrew
    Alpine.............................4
    Kinabalu..........................17
spider, happy face...................20
tarsier..............................17
vertical movement.....................6
Virunga Volcanoes..................7–10
wall creeper..........................4
weasel................................5
wolf, gray...........................12
woodland.............................10